Particulates, Coronaviruses and Greenhouse Gases

Table of Contents

Paperback 112 pages

You Tube/Peter Holst MD

Preface

Part 1 describes how diseases and global warming result from the exploitation of livestock and nature.

For half a century Mother Earth has given man dominion over his own reproduction and the reproduction of all animals and plants on earth. Meat and table eggs are still exclusively produced by artificial insemination and with breeding and incubators. African swine fever, annual flu viruses and coronavirus pandemics result from diseases in animals that can also spread to humans.

Suspended moisture droplets (aerosol) are 0.5-10 micrometers (μm) long. A person who ingests, inhales, or is otherwise exposed to positive fluid droplets contaminated with bacteria or viruses may be exposed to hundreds or thousands of bacteria or virus particles, increasing the risk of infection.

Respiratory droplets can be transmitted through coughing, sneezing, contact with contaminated surfaces or through inhaled suspended respiratory droplets (aerosol). Therefore, each individual must take appropriate measures to reduce their own exposure to these particles if there is a risk of contamination.

The accompanying table gives an overview of suspended particles in the air

µm = micron = 1/1000 millimeter	Largest diameter	Smallest diameter
Pollen granules	60 µm	12 µm
Bacteria	50 µm	0,1 µm
Suspended dust particles	10 µm	0,0006 µm
Suspended moisture droplets (aerosol)	10 µm	0,5 µm
Most mold spores	5 µm	1 µm
Legionella bacteria	5 µm	0,4 µm
Staphylococcal skin bacteria	3 µm	0,4 µm
Coli intestinal bacteria	3 µm	0,4 µm
Cigarette smoke when inhaled	1 µm	0,1 µm
Smoke from a burning cigarette	0,1 µm	0,01 µm
Hygroscopic exhalation smoke (after 2-3 sec in the airways)	3 µm	2 µm
Blue, brown asbestos	2 µm	< 0,1 µm
White asbestos	0,02 µm	< 0,01 µm
Chlamydia spores	0,3 µm	0,3 µm
Viruses	0,8 µm	0,01 µm
Coronaviruses (RNA viruses)	0,012 µm	0,009 µm

Health officials around the world agree that wearing masks can prevent the spread of virus between individuals. Certain masks are considered much more effective at minimizing the risk of exposure, especially the N95 masks.

While N95 masks from different manufacturers may have slightly different specifications, the protective properties of N95 masks are largely attributed to the ability of these masks to remove at least 95% of all particles with an average diameter of 0.3 μm or less.

Part 2 describes how to protect the earth from over-consumption and depletion. The abolition of slavery for consumption animals and optimal use of the abundance of energy that the sun gives us are indispensable in this regard. Development aid must go hand in hand with aid to reduce overpopulation in parts of the world with extreme population growth.

Part three describes how we can protect ourselves against diseases that pass from poultry and other livestock to humans and how we can grow old in a healthy way.

People who consume more animal protein have fewer antibodies, even with a small amount of animal protein. The elderly, in particular, develop diseases resulting from diet that reduces the immune response and the formation of antibodies.

Part One – Infections Transmitted from Livestock to Humans

Mass consumption of cheap burgers and broiler chickens

Viruses and bacteria went to war together

The 1918-1919 Spanish flu was a viral and bacterial pandemic. The Influenza A (avian) virus caused a flu epidemic in Fort Riley, Kansas, USA. In this fortress they raised chickens and pigs for the soldiers. A cook can be infected with the avian virus. Mutation allowed the virus to cause infection from person to person. Influenza virus (H1N1) has been transmitted to Europe via the troop transports of WWI, killing millions of people. However, most deaths from the 1918-1919 flu pandemic were due to secondary pneumonia caused by common bacteria in the upper respiratory tract. Data from the subsequent 1957 and 1968 pandemics are consistent with these findings.

Morens DM, Taubenberger JK, Fauci AS. Predominant Role of Bacterial Pneumonia as a Cause of Death in Pandemic Influenza: Implications for Pandemic Influenza Preparedness. J Infect Dis. 2008; 198 (7): 962–70

After the abolition of slavery after WWII, the trade in exotic animals and birds, parrots and songbirds has become the new business model. As a result, bird flu and contamination with bacteria such as Chlamydia pneumoniae in the human respiratory tract. Man is used as host.

More virus pandemics

Another pandemic (WHO 1980) was the HIV-1 virus pandemic resulting from the trade, sale and consumption of chimpanzee meat from the jungle. Since then, HIV/AIDS has resulted in an estimated 65 million infections and 25 million deaths. Africa has the highest number of infections.

This was followed by the Ebola virus pandemic, partly due to the consumption of bushmeat and dried bats.

Avian (ALV) and Bovine Leukemia Viruses (BLV) in our food chain

The spread of these viruses is responsible for the recent increase in colorectal and breast cancers (more on this in the relevant chapter). ALV and BLV viruses use human cells as hosts to reproduce. Since the mid-20th century, there have been an increasing number of mega-farms where pigs, cows and rabbits are bred exclusively by artificial insemination.

Viruses spread from Wet Markets

Influenza viruses and coronaviruses are mainly spread year after year from chicken farms, pig fattening farms and Wet Markets in Southeast Asia, where animals are slaughtered in the market and traded alive.

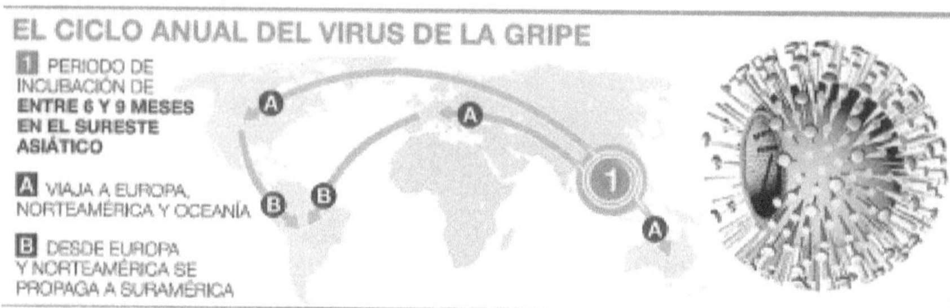

Annual Cycle of Flu Viruses

In March 2019 we returned from a cruise in Southeast Asia, the spice route, and spent a few days in Guangzhou (Canton). On arrival at the airport here, our temperature was taken, a woman with a fever was discovered and placed in quarantine. Years before the outbreak of COVID-19, since the SARS epidemic, temperature measurements and masks were already used in Southeast Asia in the fight against Coronaviruses!

Coronaviruses spread like nail bombs in humans and cause many deaths from pneumonia. Bats and rodents are carriers of these diseases. While rats and mice used to transmit disease, flying rats (bats) are now the cause of this coronavirus pandemic, which comes from wild animals in markets where the animals are traded aliv

It was not until the mid-twentieth century that life on Earth got the reproductive processes under control

Modern humans are the first living beings to gain control over their own reproduction. This was followed by control over animal reproduction through the widespread use of artificial insemination on livestock

- **Meat and eggs for consumption in factory farming are produced exclusively by artificial insemination or using incubators.**

Coronavirus, African Swine Fever, Bovine (bovine) Leukemia Virus and Avian Leukemia Virus result from diseases in animals that can also be transmitted to humans. More than 300 million farm animals in the EU spend their entire lives in a cage. The coronavirus pandemic and global lockdown have shown how vulnerable society really is.

Meat and dairy consumption continues to rise worldwide, wiping out the jungle and bringing us into contact with potentially dangerous viruses. The earth has shaped man and we are therefore indebted to our natural environment.

Viruses and bacteria teach us humans to be careful when dealing with our fellow mammals. If we do not do this or insufficiently, we will experience a lot of damage, fear and sadness as we have already experienced in the Corona year 2020. If we remain stubborn and negligent, we will pay dearly as a species and as individuals. Then we can finally be defeated by the smallest bacteria and viruses!

Humanity has increasingly come to regard the earth as its exclusive property. Including an increasing part of the jungle. We see how the world's population grew and grew, and how hardly anyone thought of sacrificing their acquired right to a daily piece of meat, resulting in a huge burden on the environment. All animals have to make do with less and less living space.

Large carnivores - such as the lion - are becoming increasingly rare in the wild, as there is less and less habitat for them and their prey thanks to humans. Worldwide, tropical forests are being cut down for soy and palm oil plantations. The soy, in turn, serves as a food source for our intensive livestock farming. As a result, on the one hand, the CO2 absorption due to the loss of tropical forests decreases and, on the other hand, soy causes an increase in CO2 via livestock farming. Our planet cannot handle the consumption of meat, there is no room for it. It is high time that people change their eating habits and stop trading exotic animals.

SARS

The SARS epidemic led to about 8,000 infections between November 2002 and July 2003. Nearly eight hundred people died. SARS has spread to more than thirty countries, including countries in Europe. Outside of China, Hong Kong, Canada, Taiwan and Singapore were hardest hit. In China, almost all provinces were affected. The Netherlands remained SARS-free. Guangdong Province (Canton) was at the center of this pandemic in 2002. The breeding, trading and eating of civet cats has been banned there since January 2004. However, recent inspections have shown that civets are still being trafficked by restaurants there, including in Hubei (Wuhan) province.

Coronaviruses and pneumonia

New viruses causing dangerous pneumonia and death in this relatively young century:

- SARS virus that emerged in China in 2002 and was spread through civet cats
- MERS virus comes from dromedaries since 2012
- SARS coronavirus II that caused the COVID 19 pandemic
- Omicron coronavirus was first detected in South Africa and has begun to spread rapidly. As mentioned earlier, Africa has the most patients treated for AIDS. Due to their reduced immune system, most lung COVID patients are here, which gives rise to extra mutations of the coronavirus and the new omicron variant.

The history of the Corona outbreak in China

It all started in 2007 with an African swine fever (ASF) pandemic when it invaded Georgia (in the Eurasian region of the Caucasus), probably caused by feeding local pigs with ASF virus-contaminated offal. including pig remains unloaded from a ship arriving from West Africa. Since then, it has spread across much of Eurasia, eventually infecting pigs and wild boars as well. The most recent further spread took place in India. The African swine fever pandemic will be worse this year (2020) than in 2019, experts say, warning that the spread of the virus, which is highly contagious and deadly to pigs, will continue.

With the global focus on the human viral pandemic of COVID-19, there is growing concern that countries are getting distracted and not focusing enough on stopping the spread of swine fever.

The ASF virus is a much 'stronger' virus than Covid-19 because it can survive for weeks and months in the environment and in processed meats. African swine fever kills nearly 100% of the animals it infects. The virus has been around for almost 100 years. in circulation, but no vaccine has yet been developed against it.

At the end of 2018, the total number of culled animals was 650,000. China's herd of pigs, by far the largest in the world, was then estimated at 360 million animals. The number of pigs had almost halved by the end of 2019 due to an epidemic of the ASF virus at the largest pork producer in the world. About 200 million pigs were culled or died as a result of the disease, reducing pork production by 40%. Due to a lack of solutions to prevent the disease and a lack of capital to raise new pigs, it can take more than 5 years for production to return to previous levels before the deadly outbreaks.

Currently, the main reservoirs of the virus are located mainly in China, Vietnam, the Philippines and much of Eastern Europe. The disease has now also spread to Papua New Guinea for the first time.

There are concerns that China is reporting too rosy data for 2020. "We see AVP here every week," said Wayne Johnson, a veterinarian at Enable Agricultural Technology Consulting agricultural company, based in Beijing.

Provincial officials are told not to file a report. China's policy has now shifted from culling to controlling and learning to live with it. The benefit that the ASF epidemic has had on the results of pig producers in China adds another interesting dimension to the story. They have already learned to live with the disease in the country and have taken advantage of the higher pork price due to the reduced supply. Profits continue to rise at the top Chinese producers such as WH Group, Wens and Muyuan. Pig producers in the US and Europe fear it is only a matter of time before the disease reaches their pig populations.

After the 2013 SARS epidemic that spread from Hong Kong, Chinese virologists previously warned that bat-borne coronaviruses would reappear to trigger the next outbreak of the disease. China is a hot spot. Bats are an incredibly diverse group that makes up a quarter of all mammals, rodents make up 50 percent and we humans make up the remaining 25% of mammals. Bats live on every continent, in close proximity to people and farms. The bats' ability to fly provides them with a wide variety of habitats, which aids in the spread of viruses. Their stools and urine can spread disease. Bats are the only flying mammals, they devour disease-causing insects by the tons, and they are essential for the pollination of many fruits, such as bananas, avocados and mangoes. Bats harbor a higher proportion of animal-to-human pathogens than any other mammal.

At the end of 2019, there was a first outbreak of a new corona virus in Wuhan. In a laboratory, research was conducted into the contamination risks of coronaviruses collected in bat caves. Have lab technicians become infected? The Corona SARS 2 virus has also passed from animals to humans in a market in the Chinese metropolis of Wuhan. In this live animal market, exotic animals such as snakes, turtles, bats, foxes and porcupines are sold for consumption. Civet cats are also still intensively bred, according to circulating price lists.

The virus is housed and spread by bats, which are boiled alive for soup in China and elsewhere in the world. It is very striking that the Year of the Rat 2020 is kicking off in China with a new coronavirus epidemic spread by bats, also known as the flying rats.

As of January 26, 2020, this COVID-19 already caused 2,751 confirmed infections with 56 deaths in China. The virus had already spread to a dozen countries. As of March 4, 2020, there were 80,409 cases, with 3,285 deaths and a spread to 86 countries.

- **Stopping the sale of wildlife in markets is essential to contain future outbreaks of animal-to-human disease.**

- ***The decision, taken by China's National People's Congress on February 24, 2020, states that illegal consumption and trafficking of wildlife will be "severely punished", as well as hunting, trading or transporting wild animals for consumption is more necessary than ever.***

Particulates

It seems like a great mystery: the corona outbreak in North Brabant. The province is increasingly developing into a hotbed for the virus in the Netherlands. Why did most people here get infected? Most Covid-19 infections were in North Brabant, the Netherlands.

A year ago we were at a wedding party near Eindhoven. I walked out with the idea of enjoying a clear evening. The air was so polluted with dust that there were no stars in sight and there was a pungent smell of neighboring pig fattening farms. How is the indoor air quality in the homes of the many pigeon fanciers, goat breeders, pig breeders and mink breeders? In a bad indoor climate, flu spreads quickly to family members and friends. What is the indoor climate like in the many pig slaughterhouses where the temperature fluctuates, work has to be done quickly in a noisy environment and people often have to yell at each other?

During the carnival in North Brabant, the corona virus entered a melting pot of visitors. The risk of contamination is particularly high in party rooms and bars.

Also in Germany and the United States, it has been proven from March to June 2020 that many workers in slaughterhouses and meat processing industries are infected with Covid-19.

The red spots on the accompanying map show more particulate matter in the air. In North Brabant this is caused by the many pig stables, goat breeders, many bird breeders and mink farms?

Particulate matter 2.5 and Covid-19 cases

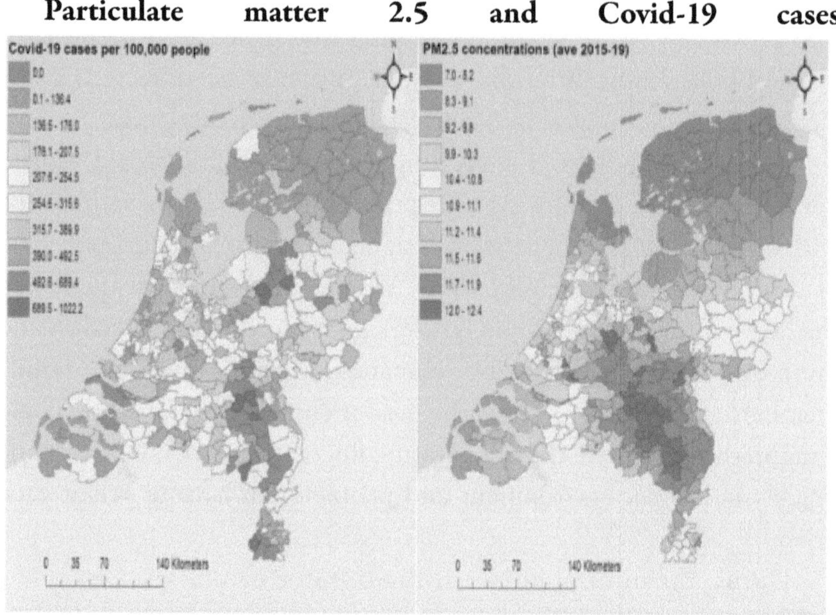

Every year in late February and early March, carnival celebrations draw thousands of people to street parties and parades – 2020 was no exception, does that explain the rapid spread of COVID-19 in North Brabant?

Due to the rural character (which is in keeping with the carnival traditions), there are many pigsties and goatherds at the carnival locations. Distribution does not even have to be done through carnival gatherings. Most of the great floats are made and parked right IN the stables of farms, and their makers frolic the day before through the mud and straw everywhere, shaking the farmer's hand and eating another sandwich there. Then hug each other during parades and also the audience on the way.

The southeastern provinces of North Brabant and Limburg are home to more than 63% of the 12 million pigs and 42% of the 101 million chickens in the Netherlands. Intensive livestock farming produces large amounts of ammonia. The concentrations are highest in air samples from the south-east of the Netherlands.

Results in the treatment of Covid-19 pneumonia

The coronavirus is a tiny little nail bomb that can enter living cells of the body and take advantage of the host cell's protein synthesis to produce offspring that allow hundreds of new virus particles to leave the host cell and spread throughout the body, resulting in a deadly disease. Several treating physicians have reported success in treating patients with fever and COVID-19 pneumonia with azithromycin. Azithromycin, a macrolide antibiotic, erythromycin and also doxycycline prevent viruses from multiplying by disrupting their protein synthesis in the host cell. These antibiotics bind to the ribosomes in the host cell, inhibit the transport of RNA and protein synthesis by viruses, thereby preventing multiplication and further spread in the body.

The majority of deaths in the 1918-1919 flu pandemic were likely the direct result of secondary bacterial pneumonia caused by common bacteria in the upper respiratory tract. Data from subsequent pandemics in 1957 and 1968 are consistent with these findings. Antibiotics also appear to have a favorable effect on the course of pneumonia in these flu epidemics. At the time of the 1918-1919 pandemic, there was no antibiotic. Alexander Fleming did not discover penicillin until 1928.

Morens DM, Taubenberger JK, Fauci AS. Predominant Role of Bacterial Pneumonia as a Cause of Death in Pandemic Influenza: Implications for Pandemic Influenza Preparedness. J Infect Dis. 2008; 198 (7): 962–70

Patients with COVID-19 have been reported to be at similar risk for secondary bacterial pneumonia, either from damage caused by the novel coronavirus itself or from invasive procedures such as intubation/ mechanical ventilation to manage respiratory failure. Prolonged immobilization in the intensive care unit has also resulted in thrombosis and emboli.

With this rapidly progressing respiratory infection, do not only take paracetamol, but also immediately in case of fever an adequate antibiotic such as azithromycin or doxycycline to prevent worse.

The long-term consequences of Covid pneumonia are insufficiently known. Does a smoker who has had a lung Covid infection with pneumonia have an extra strongly increased risk of lung cancer after ten years?

Q fever epidemic

Manure and straw from the goats are distributed over the land by farmers as fertilizer. Infected pregnant dairy goats spread spores of the bacteria with the afterbirth and amniotic fluid in the straw after delivery. As a result, the bacteria and spores of Coxiella burnetii spread through the air and infect the local population in North Brabant, even after inhaling 100 germs.

In 2007 the first outbreak of Q fever in humans took place in the Netherlands. In Northeast Brabant, 73 patients with Q fever were observed around various dairy goat farms that were infected with Coxiella burnetii. In 2008 another Q fever epidemic occurred in a larger area than in 2007. With 906 confirmed patients, this is the largest recorded Q fever epidemic in the world. In 2009 it became clear that the disease had spread over a large part of the Netherlands, almost 2,300 new infections were detected. According to the official count up to March 2010, ten people have died from a chronic form of the disease. At the peak of the Q fever outbreak in 2009-2010, 90 farms were infected. Infected animals can transmit Q fever to humans.

The bacteria end up in the environment because infected animals (which themselves do not have to show symptoms) secrete bacteria. They do this through bodily fluids, such as tears, urine, mucus, saliva, milk and amniotic fluid. A lot of bacteria are released, especially during calving or lambing. Especially with an abortion. People are infected by inhaling contaminated dust particles. The severity of the infection depends on the number of germs inhaled. In the open air, the bacteria spread as spores and can survive in this form for a long time. After inhalation, the bacteria can be absorbed into the bloodstream through the lungs and spread further in the body, in some cases leading to chronic illness. The bacteria can only multiply in living cells. Q fever does not pass from person to person.

Significantly elevated dust levels are also measured in households where birds are kept. Particulate matter with a diameter of 2.5 microns or less is the most important health risk from (indoor) air pollution. The number of particles of about 2 microns is increasing in households that keep birds. Due to excessive mucus production, smokers have more alveoli drainage than non-smokers. This transfers the dust and antigen load from the alveoli to the smaller windpipes near the bird holder that smokes.

The smaller windpipes are the preferred site of lung cancer

Both smoking and keeping birds are ultimately responsible for the malfunctioning of the "lung cleansing service" and a deficiency of immune proteins. The result is less protection of the lung mucosa cells against constant allergen and fine material that deposits on the thin mucous layer of the smaller trachea. Most lung tumors arise in the smaller windpipes, some distance from the alveoli, where initially gases and dust particles circulate.

The impact of air pollution on health.

Smog can carry large amounts of toxic and harmful substances and penetrate deep into the lungs and circulatory system through the respiratory tract, affecting human health. The 2010 Global Burden of Disease Study ranked particulate matter as the 8th leading cause of death in the world when looking at estimated deaths attributable to the independent effects of 67 risk factors. In China, this ranking could rise to 4th, accounting for about 1.2 million premature deaths in 2010. The Big Smog of 1952 in London caused about 12,000 deaths in five days. Coal fired in stoves and factories was the cause.

PM2.5, defined as particulate matter with an aerodynamic diameter of 2.5 micrometers or less, is the main health hazard from smog.

PM 2.5, due to its small size, penetrates deep into the lungs through the airways and irritates and corrodes the alveolar wall. The resulting decreased lung function will manifest itself in coughing, wheezing, difficulty breathing and other symptoms, and increase the risk of bronchial asthma, chronic obstructive pulmonary disease (COPD), emphysema and other respiratory diseases.

Lung cancer epidemic in Netherlands, Belgium and the UK

Country	Packs of cigarettes of 20 a year, in 1970	Lung cancer mortality 1984 (CBS Netherlands) 2010 (EUROSTAT)	
Italy	84	77	73
Norway	88	43	71
France	92	65	87
Finland	93	87	73
Netherlands	**108**	**117**	**108**
Belgium	**119**	**119**	**115**
West Germany	125	73	West & East Germany 79
Japan	*141*	*43*	
United Kingdom	**153**	**100**	**82**
USA	*184*	*84*	

Age-standardized lung cancer mortality (ICD 162 per 100,000 men per year) in ten different countries in 1984 and 2010 relative to the consumption of manufactured/rolled cigarettes per adult in 1970.

• **In Japan and the USA, smoking has always been much higher and the death rates from lung cancer have been much lower.**

In 2012, cancer was the cause of 31% of all deaths in the Netherlands (Eurostat). Today, about half of all men and a third of all women develop cancer and about 20% of all deaths are due to cancer. This is an impressive increase and seems to indicate that the increase in cancer is a recent biological event

In 2019, 14,000 people got lung cancer

Only 3,000 of these will be alive by 2024 (19% five-year survival rate). Lung cancer is not contagious, but year in and year out demands a lot from health care. Every day 128 people die of cancer in the Netherlands. After a diagnosis of colon cancer, the chance of being alive after five years is about 60%. For breast cancer, the chance of being alive five years after diagnosis is about 85%, much more favorable than for lung cancer.

The Netherlands, Belgium and the United Kingdom have the highest lung cancer mortality of all countries in Europe.

Besides smoking, the hobby and the breeding of tropical birds are the cause of this excess mortality.

• More lung cancer in the Netherlands, Belgium and the United Kingdom. These three countries have the largest share in the international trade and import of tropical birds via Amsterdam Schiphol, Brussels Zaventem and London Heathrow respectively.

Transport tropical birds via Amsterdam Schiphol

Bird exhibits and bird breeders fueled an explosive growth of this popular hobby

Since the slave trade and slavery were abolished 150 years ago, international trade in tropical pet animals, international human trafficking, arms industry and drug trade became the most profitable forms of trade. Worldwide, an estimated 40,000 primates, 4 million exotic birds, 640,000 reptiles and 350 million tropical fish are traded live each year. The exotics trade is estimated to be a $6 billion industry.

Keeping and breeding tropical birds is mainly a hobby of young families. The ratio between breeders and the total number of bird keepers is approximately 1:6. The degree of organization of the large bird breeders in the Netherlands is largely due to participation in competitions. Public shows, held several times a year, made the hobby more popular in the twentieth century. When pigeons are kept together with tropical birds, Chlamydia infections are more common.

In 1984 there were 7.5 million birds in households in the Netherlands. The American Veterinary Medicine Association (AVMA) counted 11-16 million companion and exotic birds in the United States in 2007. In France, 6 million pet birds were owned by households in 2010. In Belgium, every bred bird must be provided with a ring with a number by which the owner can identify the breeder. In 2011, the Ornithology Association de Belgique (AOB) registered 249 ornithological associations.

Exotic birds such as larger parrots, macaws or cockatoos are traded legally or illegally from Asia or South America.

Tropical birds were brought to the Old World late in history. After the colonization of South America and the Caribbean, increasing numbers were taken as trophies. Only with the increase in shipping and air freight could tropical birds be easily imported and traded in Europe. The largest epidemic of psittacosis occurred in 1929-1930 after the importation of infected parrots from Argentina to Europe. Hundreds of people became seriously ill and 20% died after acute fulminant illness. Initially, strict import restrictions were imposed in many countries. Not much later, parrots were again massively imported.

Bird shows and bird breeders fueled an explosive growth of this popular pastime. The housing of tropical birds in Western Europe has led to the adaptation of the psittacosis "virus", first in the flocks of the pigeon breeders who often also kept tropical birds. In many tropical bird breeders, the 'psittacosis virus' adapted and the disease that occurred in humans was less severe. The Chlamydia pneumoniae has adapted in Western Europe and the epidemic of bird flu (ornithosis) and Chlamydia pneumoniae were the result. Chlamydia pneumonia has been adapted so that this microorganism now also passes from person to person via the respiratory tract and is now so widespread in society that 98% is infected. Repeated infections with Chlamydia, which mainly occur in bird breeders and bird keepers, cause chronic respiratory diseases and lung cancer in humans.

From 1-4-1985 to 1-1-1987, 49 patients with lung cancer were younger than 65 years of age in the study in hospitals in The Hague. Of these, 21 (43%) had small cell lung cancer. Of these patients, 14 (67%) had birds in their home. The small cell form of lung cancer has the lowest chance of survival. The five-year survival rate is only 8%.

Why so much lung cancer in North Brabant?

Smoking is dangerous, but more dangerous in Belgium, the Netherlands and the United Kingdom. It is especially dangerous to smoke in the east of North Brabant near Tilburg. Is this due to the carnival or is this due to the higher dust and bacteria content of the air at the many pig fattening farms and indoors at the many bird breeders?

Organized bird breeders per 1-1-1984 per province

North Brabant 23.009
South-Holland 15.612
Gelderland 14.638
Limburg 12.426
Overijssel 10.096
North-Holland 8.973
Utrecht 6.966
Friesland 5.223
Zeeland 5.046
Groningen 4.952
Drenthe 3.726
The Netherlands 110.667

Most bird associations and organized bird keepers can be found in Limburg and North Brabant, especially in Tilburg. The number of organized bird breeders of the Dutch Bird Lovers Association grew from 1,400 in 1940 to 45,800 in 1984.

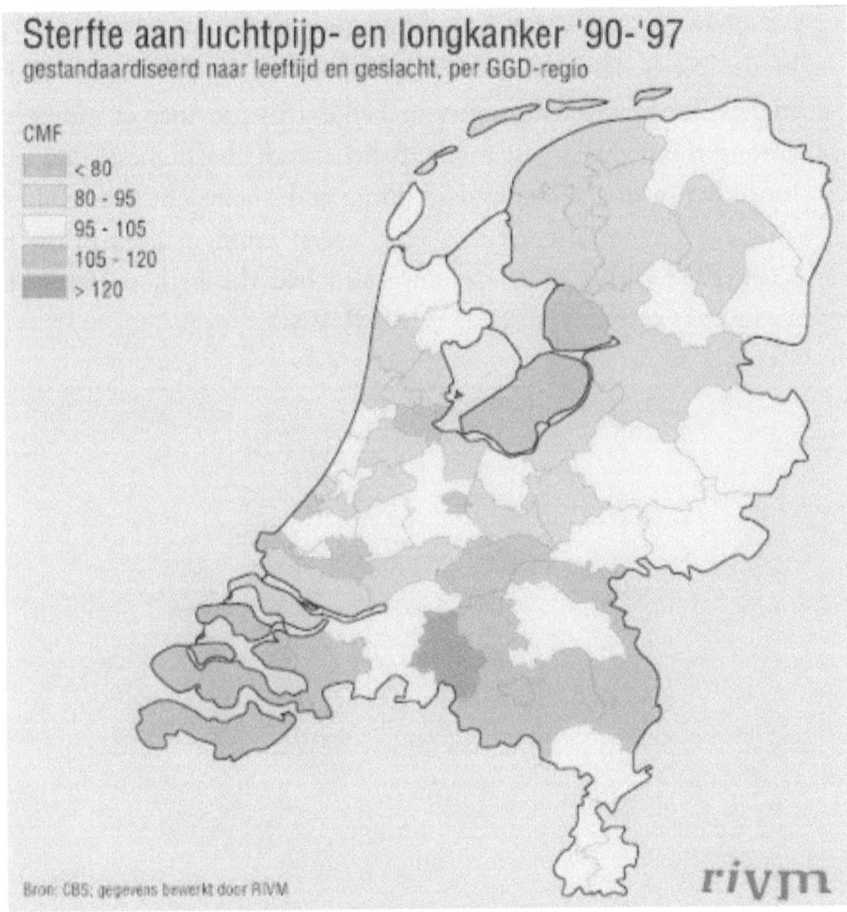

Deaths from trachea and lung cancer '90-'97, standardized by age and sex, per GGD region

There are regional differences in lung cancer mortality among men in the Netherlands. The rural province of North Brabant had the highest lung cancer mortality in 1984. This province has three cities among the eight largest municipalities with the highest lung cancer mortality in the Netherlands. Tilburg and Maastricht had the highest lung cancer mortality of the eight largest municipalities in the years 1969-1978 (CBS 1980). Tilburg also had the highest mortality in 1984 and from 1990 to 1997 (CBS, RIVM).

Lung cancer mortality (per 100,000 men per year) in various Dutch provinces and large municipalities

	1969-1978	**1984**
North-Brabant	91	114
Tilburg	**117**	**131**
Eindhoven	98	121
Breda	98	117
Utrecht	89	109
Limburg	93	104
Maastricht	**103**	**92**
Overijssel	75	104
North-Holland	95	104
Haarlem	101	114
Amsterdam	100	104
South-Holland	92	103
Rotterdam	102	123
Leiden	99	102
Gelderland	77	102
Groningen	67	90
Friesland	62	89
Drenthe	60	88
Zeeland	60	86

From 1990 to 1997, Noord Brabant again had the highest lung cancer mortality in the Netherlands, and Tilburg again.

Until 1950, the numbers of lung cancer were quite low in the Netherlands.

In 1950, 1,179 men and 167 women died from this disease. By 1983, the number had risen to 7,104 men and 800 women.

Chlamydia pneumonia (Cpn) causes lung cancer

Long-term studies with cigarette smoking machines, in hamsters, dogs and monkeys, did not show a statistically significant increase in malignant tumors in the airways, although very long exposure and high doses of smoke were used (Coggins CR 2001). These inhalation studies were performed without additional respiratory tract infection of the experimental animals. The tobacco industry has long cited the studies as evidence of no increase in lung cancer from smoking. An animal model for lung cancer was developed by repeated injection of Chlamydia pneumonia bacteria into the respiratory tract in laboratory rats, with or without the most carcinogenic component of the cigarette benzo(a)pyrene (Chu DJ 2012). Due to the combination of benzo(a)pyrene and the bacterium of the tropical bird flu in the spray, 44% of the laboratory rats developed lung cancer. The combined factors of smoking and chronic Cpn infection have effects on each other and lead to a greatly increased risk of lung cancer.

Chu DJ, Guo SG, Pan CF, Wang J, Du Y, Lu XF, Yu ZY (2012) An experimental model for induction of lung cancer in rats by Chlamydia pneumoniae. Asian Pac J Cancer Prev. 2012; 13 (6): 2819-22

Chlamydia pneumonia and lung cancer (Laurila 1997; Koyi 1999; Jackson 2000)

Laurila et al. found chronic Chlamydia pneumonia infection in 52% of lung cancer patients (n=230) and in 45% of controls. The incidence was especially increased in men younger than 60 years (3 times more often than in the control group), but not in men older than 60 years.

Jackson et al. found an increased risk of lung cancer in subjects younger than 60 years with a previous infection with Chlamydia pneumonia, but not in older subjects. (NB: Breeding tropical birds is mainly a hobby of young families)

Results with the treatment of malignant lymphomas

Malignant lymphomas were treated with tetracycline (doxycycline) and their disappearance was accompanied by the eradication of the Chlamydia pneumonia bacteria detected in the cells (**Ferreri AJ 2006**).

Chlamydia psittaci (Cp), the bacterium of psittacosis, has often been identified in malignant lymphomas. The bacterium has been detected in the tumor tissue, extracted and cell cultures made. Treatment with doxycycline (tetracycline) cures malignant lymphomas (**Ferreri AJ 2012**). Treatment with doxycycline (100 mg twice daily) for six months resolved the malignant lymphomas in 64% of patients.

The Chlamydia bacterium has no cell wall and, like viruses, is completely dependent on mucous membrane cells in the respiratory tract. By entering a body cell, the bacteria come to life. Then it copies itself. Eventually, hundreds of new bacteria leave the cell and spread further into the respiratory tract and the body. Once inside the host cell, Chlamydia is not sensitive to penicillin. Tetracycline enters the infected cells of the body by diffusion through membrane pores. Once inside the cell, tetracyclines and doxycycline inhibit internal cell metabolism, DNA and protein synthesis, preventing the Chlamydia cell parasites from producing new proteins for growth and multiplication.

Huge increase in meat production in the West

The increase in meat and dairy production in the West could only be achieved with artificial insemination of mammals and the unilateral fattening of the animals with soya meal, maize and fish meal.

Artificial insemination of pigs in intensive livestock farming

- The unrestrained breeding of animals, through artificial insemination of cattle and with poultry incubators, has a devastating effect on our health, nature and climate.
- Fast food, unnatural food and meat consumption lead to obesity, vitamin deficiencies, chronic diseases and premature death.
- Cancer is now the leading cause of premature death.

Mammals and poultry for slaughter	
	Pigs
	Suckling pigs
	Rabbits
	Cows
	Calves
	Goats
	Lambs
	Laying hens
	Broiler chickens
	The civet cat
	Minks
	Bats
	Mountain marmots
	Chimpanzees
	Dromedaries
	Wildlife
	Whales

Humans take up all the space on earth, most animals for slaughter are artificially fertilized and fattened in stables or cages.

Before 1950, the bullrunner still visited the farms with a bull to fertilize the cows.

What are the risks of higher meat production?

Increased mortality from brain tumors has been observed in veterinarians (**Blair A**). Vets and K.I. assistants do a lot of vaginal internal examinations in cows and become infected with the bovine leukemia virus through aerosol formation. Transmission of the bovine leukemia virus (BLV) through the uterus and birth canal during childbirth plays a critical role in the spread and persistence of BLV infection in cattle.

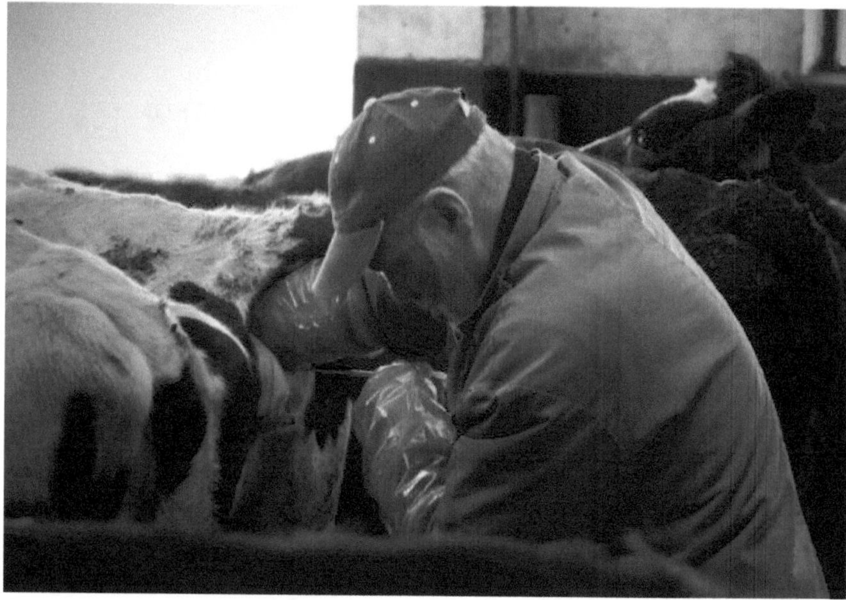

Veterinarians and AI employees come into contact with the bovine leukemia virus (BLV), a cancer-causing virus, in their work. Note: this vet does not wear a mouth cap! Bovine leukemia is an economically important infection of dairy cattle worldwide.

The incidence of infections in the Canadian dairy herd is high and increasing. Seventy percent of the herds were identified as BLV positive (one or more positive animals).

Nekouei O, VanLeeuwen J, Sanchez J, Kelton D, Tiwari A, Keefe G Herd-level risk factors for infection with bovine leukemia virus in Canadian dairy herds. Prev Vet Med. 2015; 119 (3-4)

The deaths of 5,016 veterinarians were examined and compared with those of the general US population. The death rates were significantly increased from malignant lymphoma and leukemia, colon, brain and skin.

Blair A, Hayes HM Jr. (1982) Mortality patterns among US veterinarians, 1947-1977: an expanded study. Int J Epidemiol. 1982 Dec;11(4):391-7.

The increased risk of esophageal, colon, brain and pancreatic cancer and melanoma among veterinarians in Sweden could not be explained by the socioeconomic status of this profession. Occupational exposure to cancer-causing viruses in livestock are possible sources.

Travier N, Gridley G, Blair A, Dosemeci M, Boffetta P. Cancer incidence among male Swedish veterinarians and other workers of the veterinary industry: a record-linkage study. Cancer Causes Control. 2003 (6):587-93.

Cows are continuously re-fertilized by artificial insemination after the calves are born so that their milk never stops flowing.

Artificial insemination (AI) in cows was introduced in Friesland in 1935. The bull's semen is frozen in "straws" and then introduced into the animal by a vet or AI assistant. Calves are taken from their mothers shortly after birth.

Their calves grow up to be dairy cows or are bred for veal. For the production of milk and cheese, the mother cow has to give birth to as many calves as possible. Milk, cheese and meat production are inextricably linked.

Raw egg proteins and raw milk products

Industrially processed foods contain a large proportion of liquid chicken egg proteins, which in some cases are not processed with sufficient heat. Eggs are beaten on crushers, yolk and white are separated, eggshells and hail wisps are removed through filters, and the egg white product is heated to 56° Celsius.

In the Netherlands (1983) 20,000 tons of liquid chicken protein was produced for industry, marginally pasteurized, sometimes undercooked. The confectioner processes a large number of products containing eggs. This can be the pasteurized egg whites, or it can be processed fresh eggs. The "whites" are collected in a special container. Housewives also sometimes come into contact with raw egg whites when making cake batter or desserts at home. Or when they whip the raw egg whites.

Raw egg proteins are processed in:

SUGAR GLAZING raw egg whites with powdered sugar

CREAM FONDANT raw proteins with butter, sugar and liqueur

OMELET SIBÉRIEN raw egg whites with sugar

BAVAROIS raw egg whites with sugar, cream, gelatin, fruit

ICE CREAM raw egg whites with sugar, milk and cream.

And in:

STEAK TARTAR with a raw egg

Raw milk products

Raw milk is milk from cows, sheep or goats that has not been pasteurized to kill harmful bacteria and viruses. Raw milk and cheese made from raw milk also contain raw proteins and can contain harmful bacteria and viruses.

Employee risks in the poultry industry

Cancer-causing viruses are found and cause tumors in chickens and turkeys. Some are carriers and spreaders of these viruses. Virus has been shown in chicken products and eggs, so exposure to humans is universal and almost inevitable. These viruses are not highly contagious, yet have the ability to infect and transform human cells. Antibodies against avian leukemia viruses (ALV) and reticuloendothelial viruses (REV) have been found in blood sera from workers in poultry slaughterhouses. Cancer mortality has been studied in 20,132 workers in poultry slaughterhouses and processing plants, a group with the highest human exposure to these viruses. Significantly increased risks were observed in poultry workers as a whole or in subgroups, for several cancers: cancer of the mouth and pharynx; pancreas; trachea / bronchus / lung; brain; cervix; lymphatic leukemia; and tumors of the blood-forming and lymphatic systems.

Metayer C, Johnson ES, Rice JC (1998) Nested case-control study of tumors of the hemopoietic and lymphatic systems among workers in the meat industry. Am J Epidemiol 147(8):727-38

Johnson ES, Ndetan H, Lo KM (2010) Cancer mortality in poultry slaughtering / processing plant workers belonging a union pension fund. Environ Res 110(6):588-94

Brain cancer is more common in poultry farmers involved in the killing of chickens. Killing chickens was associated with a nearly sixfold increase in brain cancer risk. Workers in poultry slaughterhouses and processing plants often process thousands of chickens every day, come into contact with poultry meat, organs and blood, and are at risk of injury that provides a pathway for viruses and other microbial agents to enter the body. They also operate in confined spaces for extended periods of time, which increases the risk of inhaling microbes. Viruses known to cause cancer in poultry may be responsible for the increased incidence of cancer in poultry farmers who kill chickens.

Gandhi S, Felini MJ, Nidetan H, Cardarelli K, Jadhav S, Faramawi M, Johnson ES (2014) A pilot case cohort study of brain cancer in poultry and control workers. Nutr Cancer.

Professor ES Johnson, former epidemiologist at the University of Fort Worth, Texas, has developed and patented the only test to date that can detect the presence of cancer-causing viruses in the genome of tumor cells of workers with these cancers.

How are animals fed in megafarms?

Southeastern Pacific anchovies are sold as animal feed to factory farms in Europe. About a third of the total catch is fed to consumption animals, mainly farmed fish, pigs and chickens. European fishermen are required to land all by-catches by 2020. In addition to by-catches, the fish processing industry also produces a significant amount of recyclable waste, such as skins, bones, fish heads and internal organs. Fish meal can be created by hydrolysis of the fish from by-catches and fish remains, which is in great demand. Especially in the fish farms in the Mediterranean.

Tuna, salmon, cattle, pigs and chickens grow faster and become fatter due to fishmeal. More profit can be made and the time to slaughter is reduced. About 20-30 million tons of fish, anchovies, herring, mackerel and sprat have been removed from the southeastern Pacific Ocean in recent decades for the production of fish oil and fish meal.

1000 years ago in Europe, population growth caused the decline of freshwater fishing and fishermen began fishing the oceans in greater numbers for the first time. Five hundred years ago, it was the decline of inshore fishing that gave rise to deep-sea trawling. Worldwide 20 billion is granted annually: 6.3 billion is spent only on subsidies for fuel, 8 billion extra goes to the maintenance of the major ports.

Small-scale fisheries use 75% less energy to catch the same amount of fish, are more environmentally friendly and employ many more people.

Mega farms with only cows, calves, pigs or chickens feed the animals with soy meal, fish meal and low doses of antibiotics to fatten the animals faster and make more profit. As a result, the animal fat content of steak, pork and chicken meat has increased dramatically.

Diseases later in life due to fast food

After the transition to a more industrialized diet, hamburgers, sugary drinks and fast food in these countries, increased consumption of energy, animal proteins, animal fats and red meat was observed in several parts of the world. With more than 225 million overweight people in 2016, the US has the largest number of overweight people in the world. The US also exports these unhealthy eating habits around the world. Poorer nutrition is the leading cause of disease and mortality in the United States. With an average life expectancy of 78.1 years, the United States is only number fifty in the world, despite being the richest nation in the world with the most advanced medical technology. The Netherlands is doing slightly better with an average life expectancy of 79.2 years, but less than most other European countries. Even despite the country's alarmingly high suicide rate, Japanese live an average of 82.1 years.

Diet-related diseases are the leading cause of death in the United States, surpassing smoking as a cause of premature death.

The number of overweight or obese people increased between 1990 and 2016. Poor diet contributed to 14 percent, while smoking accounted for 11 percent. Obesity and high blood pressure accounted for 11 and 8 percent, respectively. The number one cause of death in America is the American diet.

When people move from low-risk to high-risk countries, their disease rates almost always change to match the new environment. New diet, new diseases. But the reverse is also true. Eating the standard American diet and switching to a diet with more plant-based foods, such as fruits and vegetables, may lower your risk.

Cancer is now the most common cause of death in Western Europe, more common than chronic obstructive pulmonary disease (COPD), cardiovascular disease and diabetes (IHD). While death rates for COPD and IHD are falling due to improved health care, cancer death rates have risen. The consumption of animal fats and proteins has increased considerably since the last century. The production of meat (products), poultry, pork and other meat tripled between 1980 and 2010 and is expected to double again by 2050. At present, 70 billion farm animals are bred for food every year.

As we get older, we notice which unhealthy lifestyle habits have taken over us. The body constantly renews itself through the ingested food and within a few years all cells and tissues are constantly completely rebuilt. With age, the choice of animal or vegetable proteins and fats in the daily diet is of great importance for protection against chronic diseases and cancer. Cardiovascular diseases, obesity and uncontrolled growth of derailed cells are the result of an excess of animal proteins and fats in the daily diet. The chicken leukemia virus and the bovine leukemia virus in our food chain are related to common cancers. The time without symptoms is 50% - 70% of the total growth of a tumor and cancer usually manifests itself later in life.

In Japan and Korea, large-scale imports of beef and pork began after World War II and after the Korean War, respectively. In 1970 in Japan and 1990 in Korea a strong increase in the number of colon cancers was observed. Consumption of fried beef (including shabu-shabu, Korean yukhoe and Japanese yukke) became very popular in both countries.

A specific meat factor, presumably one or more thermoresistant bovine cancer-causing viruses (eg polyoma, papilloma or single-stranded DNA viruses), can contaminate beef and lead to long-lasting infections in the intestinal tract.

Zur Hausen H (2012) Red meat consumption and cancer: reasons for suspect involvement or bovine infectious factors in colorectal cancer. Int J Cancer. 2012 Jun 1; 130(11): 2475-83

In East Asia, there has been an increased consumption of energy, animal fat and red meat in recent decades. Breast, colon, prostate, esophageal and gastric cancer mortality data for China (1988-2000), Hong Kong (1960-2006), Japan (1950-2006), Korea (1985-2006), and Singapore (1963-2006) were obtained from the WHO. A noticeable increase in breast, colon and prostate cancer death rates and a decrease in esophageal and gastric cancers was observed in the selected countries during the study periods. For example, the annual percentage increase in breast cancer deaths from 1985 to 1993 in Korea was 5.5% and prostate cancer deaths rose by 3.2% per year from 1958 to 1993 in Japan.

These changes in cancer mortality followed about 10 years after the transition to more industrially prepared foods, hamburgers, sugary drinks and fast foods in these countries.

Zhang J, Dhakai IB, Zhao Z, Li L Trends in mortality from cancers of the breast, colon, prostate, esophagus, and stomach in East Asia: role of nutrition transition. Eur J Cancer Prev 2012 Sep; 21 (5): 480-9

Prostate cancer death rates rose dramatically (25x) in Japan after World War II. After the war, milk consumption increased 20x, meat 9x and eggs 7x. Milk contains high amounts of estrogens plus protein and saturated fats. The recent increase in its use is likely the cause of the increase in prostate cancer in Japan.

Ganmaa D, Li XM, Qin LQ et al. The experience of Japan as a clue to the etiology of testicular and prostatic cancers. Med Hypotheses. 2003 May; 60 (5): 724-30

The inhabitants of Tuvalu, Fiji, Samoa and the Cook Islands are massively overweight. According to the World Health Organization (WHO), nine of the ten fattest countries in the world belong to the Pacific Islands. Tonga (4th, 90.8%), Samoa (6th, 80.4%) and the US (9th, 74.1%). In some countries, up to 95 percent of the adult population is overweight. The number of obese people, extremely overweight, ranges from 35 to 50 percent.

The Cook Islands (90.9% overweight) are in third place in the world ranking. Just over half of the population is obese. Inexpensive factory-processed foods have replaced the original diet of fresh fish and vegetables. Fresh fish is relatively expensive, with this money you can buy several hamburger meals. A bottle of Coke is cheaper here than a bottle of water. 75% of the people on Samoa are extremely overweight.

Recent increase in cancer

Colon cancer

An increased risk of colon cancer has been shown with the consumption of undercooked red meat. Beef and pork were widely imported into Japan and Korea after World War II and the Korean War. After 1970 a sharp increase in the number of patients with colon cancer was observed in Japan and after 1990 in Korea. The consumption of undercooked beef (eg Shabu-shabu, Korean Yukhoe and Japanese Yukke) became very popular in both countries.

Virus as a source of colon cancer

A specific beef factor, probably a heat-resistant carcinogenic bovine virus (DNA virus: e.g. polyoma and papilloma viruses and RNA virus: bovine leukemia virus BLV) can contaminate beef and cause long-lasting intestinal infections after human consumption (Zur Hausen H 2012).

Polyoma viruses in burgers

Ground beef samples have shown three types of polyomavirus, which are resistant to barbecue temperatures and are carcinogenic to their natural hosts. Papilloma and polyoma viruses in particular are resistant to moderately heated steak tartare, where central parts of the meat are not heated above 40 - 70 degrees Celsius. These viruses tolerate 80 degrees Celsius for 30 minutes without losing their ability to cause infections. These viruses are also insufficiently inactivated during the pasteurization of dairy products (Peretti A. 2015).

Peretti A, FitzGerald PC, Bliskovsky V, Buck CB, Pastrana DV Hamburger polyomaviruses. J Gen Virol 2015 Apr;96(Pt 4):833-9 https://www.ncbi.nlm.nih.gov/pubmed/25568187

Zhang J, Dhakai IB, Zhao Z, Li L Trends in mortality from cancers of the breast, colon, prostate, esophagus, and stomach in East Asia: role of nutrition transition. Eur J Cancer Prev 2012 Sep;21(5):480-9

Zur Hausen H (2012) Red meat consumption and cancer: reasons to suspect involvement of bovine infectious factors in colorectal cancer. Int J Cancer 2012, 130(11):2475-83

Breast cancer

Humans are exposed to cancer-causing viruses commonly found in animals in the food chain, such as layers, eggs, broilers and dairy cows. The avian leukemia virus (ALV) and bovine leukemia viruses (BLV) are RNA viruses and have been shown in breast cancer cells.

Bovine Leukemia Virus (BLV) has been demonstrated in breast cancer cells

Milk, dairy products and eggs are probably the cause (Buehring). Consumption of unpasteurized dairy products, or cheese made from raw milk, or undercooked beef on the BBQ can transmit this contagious virus to humans.

Buehring GC (2015) has shown that 39% of people in a San Francisco Bay Area have antibodies to BLV in their blood, indicative of exposure to BLV. Almost all cow's milk contains BLV bovine leukemia virus. In a study of 213 women, BLV-related DNA was found in breast tissue from women diagnosed with breast cancer, not in breast tissue from women with no history of breast cancer.

Buehring GC, Shen HM, Jensen HM, Jin DL, Hudes M, Block G (2015) Exposure to Bovine Leukemia Virus Is Associated with Breast Cancer: A Case-Control Study. 2015 Sep. 2;10(9):e0134304

https://www.ncbi.nlm.nih.gov/pubmed/26332838

Avian Leukemia Virus (ALV) is a retrovirus that infects large parts of modern poultry farming and causes considerable economic damage. The virus is present in chickens and eggs. Transmission by infected chickens of viruses to the eggs; processing raw, undercooked protein in food; this is how the ALV virus reaches humans **(Pham TD).**

How great is the loss of animal and plant species?

- In the past 50 years, homo sapiens have wiped out 60% of mammals, birds, fish and reptiles in the wild
- Mankind has destroyed 83% of all mammals and half of all plants since the dawn of civilization.
- Wildlife hunting in tropical forests reduces bird and mammal populations.

Our ancestors probably consumed bushmeat, wild animals that were killed for food. In the 20th century, however, commercial hunting with firearms and wire snares for the supply of logging and oil exploration concessions along new road networks dramatically increased catches in Central African forests. An estimated 579 million wild animals are captured and consumed annually in the Congo Basin, equivalent to 4.5 million tons of meat, with the addition of possibly 5 million tons of Amazonian mammal meat. Tropical lowland forests contain the world's greatest biodiversity and therefore harbor a reservoir of germs that can be transmitted from animals to humans.

The wildlife trade in general generates more than a billion direct and indirect contacts between humans and domestic animals each year. The wide range of tissue and fluid exposures associated with the bushmeat industry's hunting and slaughter can make these interactions with wildlife particularly risky. In Africa, as many as 30 different primate species are hunted and processed by the bushmeat industry.

- **One billion people suffer from hunger, while 70 billion animals are fattened and eaten every year.**

We can no longer ignore the impact of today's unsustainable production models.

Benítez-López A, Alkemade R, Schipper AM et al. The impact of hunting on tropical mammal and bird populations. Science 2017 356(6334):180-183

Population growth to seven, eight or nine billion causes food shortages, diseases and climate change due to intensive meat production. The large number of animals locked up for meat consumption is the perfect system as a source for pathogenic viruses such as highly pathogenic influenza, corona virus and leukemia virus. Humanity is moving in the same direction as the species we have seen disappear.

How do we make a greenhouse of the earth?

- Each year, Alaska's glaciers shorten hundreds of feet as the ice melts.
- In the US, cattle emit about 5.5 million m3 of methane, a gas 25 times more potent than CO_2.

Livestock is responsible for at least 14.5% and according to some studies as much as 51% of man-made greenhouse gases.

CE Delft, Fraunhofer Institute for Systems and Innovation Research and LEI Wageningen. Behavioral climate change mitigation options and their appropriate inclusion in quantitative longer-term policy scenarios. Delft, January 2012

All life depends on the oceans. Circulation in the North Atlantic has slowed to its lowest level in centuries. The slowdown in the Gulf Stream is destroying fisheries and will lead to sea level rise. Every year, large amounts of nitrogen from fertilizers and manure are distributed over agricultural land. No doubt it increases crop yield, but plants don't take it up completely, so more fertilizer and animal waste is added than the plants need. Only a fraction of what is applied to the soil ends up in the crops. The rest flows to our rivers. Nitrogen and phosphorus levels, dead organisms are increasing in the Gulf of Mexico, Rhone Delta, North Sea, Baltic and Adriatic Seas. The oxygen content in these coastal waters is falling. (**Phillip Lymbery 2017**).

How did we make a desert of the land on earth?

Allan Savory

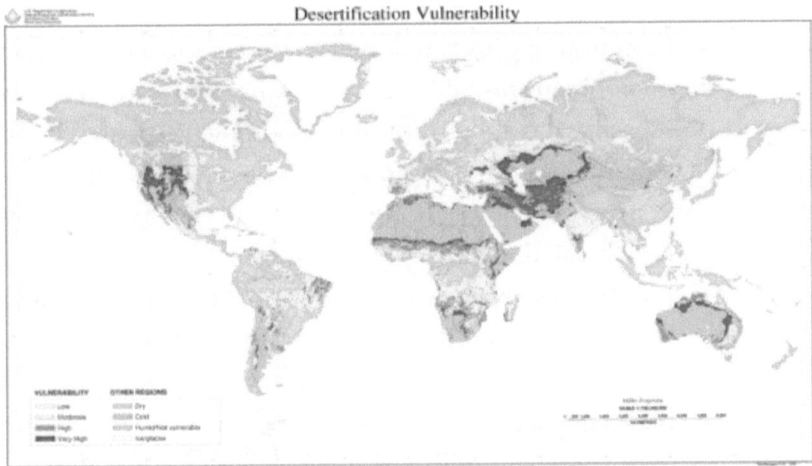

Desertification on 2/3 of the land (grey, yellow and red areas)

- When humans mastered fire and language and developed weapons such as spears and axes, they were formidable predators. This was especially true in the grasslands where their prey roamed in herds. The grasslands with their deep, watery and carbonaceous soils had evolved over millions of years thanks to the balance between grazing animals and the predators that fed on them.

- Modern farming practices contribute significantly to desertification and climate change due to water and air pollution from agriculture and intensive rearing of pigs, poultry and livestock. By breaking down farmland, we are reducing its tremendous capacity to hold and retain carbon.
- Chemical fertilizers to increase production have killed soil microorganisms, reduced soil fertility and water retention capacity, and led to additional flooding. Pesticides used to treat internal parasites in animals have resulted in the destruction of dung beetles, which are vital for soil renewal.
- Fires break down the ground cover in such a way that it is easily carried away by rain and wind. Huge man-made deserts have been created. According to NASA photos from space, about two-thirds of the country has turned into desert.
- Banks finance farmers to rent out their pastures for solar panels and windmills and to set up even more megafarms using electrical energy.
- Ever-growing meat production is causing drought and hunger in large parts of the world. Fast food and an increase in meat consumption in the West are also being imitated in other parts of the world.

Part Two - Save Ourselves And The Earth From Flooding

S. O. S.

Save Our Planet Earth

After the Spanish flu (Influenza H1N1 pandemic, originating from the poultry industry), which reigned from 1918 to 1920, the economic situation was bad due to pandemic measures, schools and resorts were closed. An economic crisis also followed in the interwar period. After World War II, the United Nations was established to define land and human rights. Later, human slavery was abolished.

Every year we protect ourselves with Influenza Avian vaccine against infections with highly pathogenic viruses from poultry farms in Southeast Asia. From now on we will also have to protect ourselves annually against the Coronaviruses from the meat industry!

Will there be an economic crisis after the umpteenth wave of Coronaviruses from 2019 to 2021 (meat industry pandemic) and Omicron 2022?

The climate crisis seems inevitable and there is a great loss of plant and animal species. Will man gain insight and abolish the slavery of food animals?

Bear with three cubs in the tundra wilderness of Denali

All mammals should have the ability to care for their young themselves. Artificial insemination of livestock and intensive farming of cows, goats, pigs, sheep and rabbits is a serious offense. All mammals have the right to live without cruel treatment and lifelong confinement.

Grasslands and grazing animals have a tremendous capacity to retain carbon. Agriculture can produce fruits and vegetables as food for humans better than the corn and soybeans with which we now fatten the animals and lock them up in factory farms to be eaten by us.

Birth control

In order to put an end to the abuses in the Kempen that doctor Ferdinand Peeters encountered in his daily practice, he went in search of a means with which the woman could regulate her own fertility. The contraceptive pill Enovid, marketed by the American biologist Gregory Pincus in 1957, still had too many side effects and was only authorized as a remedy for painful periods. In 1959, Dr. Peeters conducted a series of clinical tests with a hormone preparation offered by the German company Schering AG from his laboratory at the Sint-Elizabeth hospital in Turnhout.

For six months, Dr. Peeters and his assistants Reimond Oeyen and Marcel Van Roy tested the preparation on fifty Kempen women for whom even more children posed a major health risk. After countless experiments to find the right (more than half-lower) dose of the two hormones (progestin and estrogen), Peeters explained the findings to Schering in Berlin in 1960. The results were astonishing. None of the women became pregnant and there were hardly any side effects.

After the preparation of Peeters (SH 639) also proved to be safe and efficient in the United States, Japan and the United Kingdom, Schering launched the Anovlar pill in January 1961.

Pincus tacitly acknowledged the superiority of Peeters' pill by halving the dose of Enovid in July 1961. Ultimately, it was Pincus who took credit and (wrongly) went down in world history as the inventor of the contraceptive pill. Fearing being robbed by the church, however, the very Catholic Peeters did not give much publicity to his invention.

Moreover, the pioneering research of a 'rural doctor' also aroused much contempt among jealous professors at the KUL, where Peeters was also active.

- Sheep gut and since 1844 Goodyear rubber has been used as a contraceptive.
- The pill, IVF and artificial insemination led to the breakthrough in the mid-twentieth century.
- The German company Schering AG launched the Anovlar pill in January 1961.
- The pill proved to be a particularly powerful emancipatory tool, both in America and Europe. The pill radically changed the balance of power between men and women - and with it society as a whole.

The discovery of the pill was a major revolution. For the first time in human history, sexual intercourse and reproduction could be technically and artificially separated. The enormous commercial success was dazzling, both for medicine and for theology. And each one tried to outdo the other by providing a good justification for birth control. Contraception caused a radical break with the lives of all previous generations and civilizations. Legal abortion has now been introduced, the number of divorces has increased spectacularly, and euthanasia has been advocated when life is no longer experienced as meaningful. The family as the basis of church and society has crumbled into all kinds of free forms of cohabitation. All forms of sexuality are openly discussed. The rise of gays, Me Too for sexual harassment, sexual harassment and rape, incest and pedophilia, transgender reassignment, genital mutilation in other cultures. The sexual abstinence of celibacy also had its problems, such as pedophilia among priests.

Yet the balance of sudden sexual freedom is generally positive, thanks to the increased emancipation of women and the awareness of sexual problems and lack of freedom that have been hidden for years. There are more and more women in leadership positions. We are waiting for the first woman as pope, now that women have been admitted to the church. In the metropolis of Guangzhou (the former canton), our guide pointed out a female Buddha in the largest temple.

What does the Sun give us?

The Namib Desert is one of the sunniest places in the world.

In 54 minutes, the sun radiates to the earth an amount of solar energy that the entire world consumes in a year. All free renewable energy. Only it's in the wrong place at the wrong time in the wrong shape. If we convert all that energy, which is now lost every day, into hydrogen, the scarcity of energy will be over.

The Namib Desert near Lüderitz, a port city on the Atlantic Ocean on the southwest coast of Namibia, is one of the sunniest places in the world.

This desert is 200 km wide and stretches 2000 km from Angola in the north to the Orange River in the south along the Atlantic Ocean. A suitable area for energy production can be found in this 81,000 km2. A combination of solar panels and hydrogen gas production with transport to sea can offer Namibia economic benefits.

Groningen can also earn money from the transition to hydrogen gas.

The Netherlands already has an extensive gas network from Groningen to the rest of the Netherlands. This natural gas network can be used for hydrogen gas transport without too many adjustments

Veendam has the first larger hydrogen power plant in the Netherlands that uses solar energy. It is an important step in Groningen's mission to become the leading hydrogen province in the Netherlands. Renewable energy from the electricity grid and 5,000 solar panels on the site supply green power to the plant, which can convert one megawatt of sustainable electricity into hydrogen. A hydrogen industry in Groningen can supply the large amounts of energy that will be lost at the close of the oil and gas era. All this energy can be stored in the form of hydrogen and transported via the gas network and in liquid form by ship.

Every day that we do nothing with the storage of solar and wind energy here and there in the Netherlands and in the deserts is a lost day. Let us not only say goodbye to fossil fuels, but also to biofuels. The demand for palm oil has increased too much due to Western policy to stimulate the use of biofuels. Fires have intensified in recent years in the Brazilian Amazon and the Indonesian tropical rainforests of Borneo and Papua New Guinea. The slash and burn method is used to cultivate natural soil for palm tree plantations. Hydrogen as a primary energy source may be the solution.

Canadian engineers have found a way to produce hydrogen from oil fields and tar sands relatively easily and cheaply.

By injecting oxygen into the tar sands, the temperature in the soil seems to rise. This releases hydrogen gas from the oil, which can be separated from other gases by special membrane filters. Even with oil fields still in use, this technique can be used. In this way, a polluting fossil raw material can be given a second life and produce the energy carrier of the future. Hydrogen production is a cost-effective alternative to energy production from oil fields and tar sands. By placing hydrogen filter membranes in the production wells, only the hydrogen is extracted and unwanted by-products such as carbon dioxide diode and methane remain in the soil.

The existing infrastructure and distribution channels around the oil fields would suffice, keeping production costs low. At the moment it costs about 2 dollars to produce a kilo of H2, but with the new method it would be only 10 to 50 cents. The required oxygen can be produced on site. This requires no more than 5 percent of the energy produced.

Researchers at the University of Waterloo in Canada have also developed a new fuel cell that will last at least ten times longer than current technology. These fuel cells will therefore be much cheaper.

A greenhouse complex in Westland wants to be able to produce energy-neutral fruit and vegetables all year round with the help of a hydrogen power plant. Farmers can repurpose their barns. With local production with more supply and diversity, the current supply of fruit and vegetables across the equator may decrease.

Part three – Stay healthy

This Mardi Gras float represents the deeper meaning of Carnival - carnem levare (Latin), omit meat and eat an abundance of fruits and vegetables.

Stay healthy for 100 years

We eat everything that is tasty. When it's cheap and tasty, it also accelerates chronic disease and tumor formation. That's how our food system works. The raw materials for factory preparation are limited. Soy, corn, eggs, refined sugars, animal proteins and trans fats. These are the main ingredients that the largest food companies use to make the foods that are widely available. It's not that these big companies don't care. In fact, it's hard for them to do anything else. In parts of the world with less chemical agriculture, less factory food processing, cancer deaths are lower. Fruits and vegetables make bioactive compounds that slow the growth and reproduction of invaders.

Plant-based food prolongs life

Voluntary starvation is unlikely to gain much popularity as a life-prolonging strategy and is risky in terms of developing shortages. Vegetable proteins - especially those from vegetables or nuts - contain less methionine than animal proteins. Several animal studies with methionine-restricted diets have shown inhibition of cancer cell growth and extended healthy lifespan in laboratory animals. American researchers have looked at nutrition data among 130,000 people for 30 years. They found a reduced risk of premature death in those who ate more vegetable protein and a higher risk in those who ate more animal protein. Each increase of 3% more vegetable protein in the diet reduced the risk of death from any cause by 10%. A 12% lower risk of death from cardiovascular disease was also shown. But a 10% higher proportion of animal protein in the diet led to a 2% higher risk of death and an 8% higher chance of dying from a heart problem **(Song M.).**

Unnatural food is the cause of deficiencies and chronic diseases. Fast food, lots of meat products and few fruits and vegetables weaken the natural defenses. How can one switch to only natural food?

Oatmeal flakes

Whole wheat is an important source of antioxidants and bioactive compounds such as phenols, flavonoids and carotenoids. Flavonoids work very well as antioxidants and prevent the conversion of calories into fat. They also have a protective effect on the blood vessels. Most of the health-promoting compounds of whole wheat are present in the germ and bran.

Vitamin C

Vitamin C is necessary for building connective tissue proteins. A defective production of these connective tissue proteins weakens the blood vessels, resulting in bleeding.

Due to the frequent use of pesticides in agriculture, the content of bioactive antibodies in fruit and vegetables has been reduced, which further weakens our defenses against cell infections. Vitamin C 1000mg protects against the common cold and the possible infections associated with the common cold. It also appeared that patients who took 2 grams of vitamin C per day spent a shorter time in the ICU. Patients who received vitamin C needed significantly less time for ventilation.

Hemila H, Chalker E Vitamin C Can Shorten the length of stay in the ICU A Meta-Analysis. Nutrients (2019) 11(4):708

Vitamin D3 (cholecalciferol) is the precursor to the powerful steroid hormone calcitriol, which has widespread effects throughout the body. Several studies have shown that vitamin D deficiency increases the risk of developing cancer and that vitamin D3 intake may be an economical and safe way to reduce cancer incidence and improve cancer treatment outcome. Sufficient vitamin D also increases bone density, reducing the risk of fractures.

Curcuma

Curcuma is the spice that turns curry powder yellow. In areas where this herb is eaten on a daily basis, a number of cancer diseases that are common in Western Europe are less common.

Multiple myeloma (Kahler's disease) is a cancer of the cells of the bone marrow. Plasma cells in the bone marrow ensure the production of antibodies against viruses and bacteria. This cancer is rare in India (blue area) and common in Western Europe (red area in attached image).

Herbs and fruits often contain the nutrients that fast food lacks. Numerous plant nutrients, one of the best known is curcuma, protect against damage to the chromosomes of cells in the body. Malignant growth generally results from a DNA attack upon exposure to chemicals, radiation, or viruses that take possession of the host cell's DNA. Curcuma protects against cell DNA damage. Infected or damaged cells are now seen as abnormal and removed from the body.

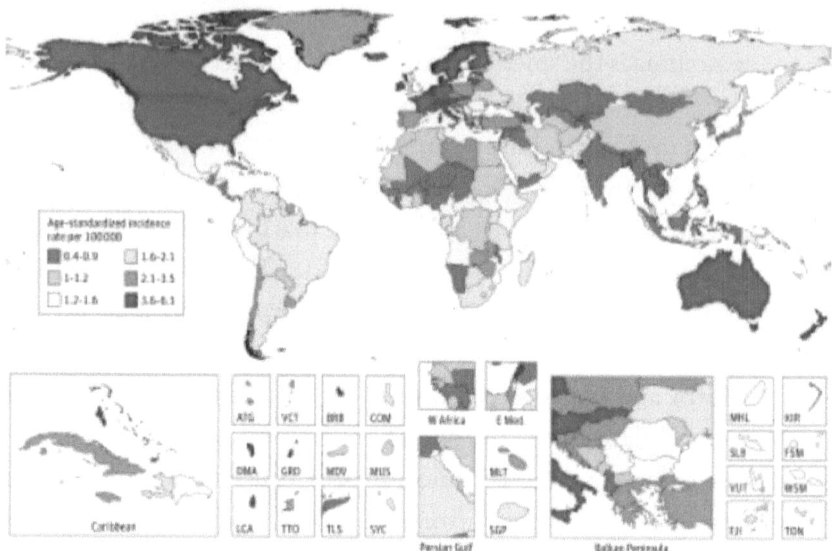

Andrew J. Cowan, Christine Allen, Aleksandra Barac et al. Global burden of multiple myeloma: A systematic analysis for the global burden of disease study. JAMA Oncol 2018 Sep 1;4(9):1221-1227

Overall cancer mortality is much lower in India than in Western countries. American men get 23 times more prostate cancer than men in India. Americans 8 and 14 times more likely to develop melanoma, 10 to 11 times more colon cancer, 9 times more uterine cancer, 7-17 times more lung cancer, 7-8 times more bladder cancer, 5 times more breast cancer, and 9 to 12 times more likely kidney cancer than in India. This is not just 5, 10 or 20 percent, but 5, 10 or 20 times more. Indians together make up one sixth of the world's population. They have the lowest cancer rates in the world. A decreased incidence of cancer cannot be the result of increased consumption of herbs alone. Several dietary factors may contribute to India's low overall cancer rates. In addition to the high consumption of spices, Indians eat little (red) meat (holy cow) and 40 percent of Indians are vegetarian. India is one of the largest producers and consumers of fresh fruits and vegetables. In addition to curcuma, Indians also eat many other herbs. Curcuma contains the highest antioxidant content (Holt PR) per unit weight.

Omega-3 fish oil capsules, EPA and DHA have a beneficial effect on the heart and brain. Human brain cells contain a lot of DHA. Sufficient fish in our menu is very important (Crawford MA 2012).

Benefits of the Mediterranean Diet

The diet in Spain, Italy and Greece is one of the healthiest eating habits in the world. Apparently there are fewer diseases here and the mortality from some chronic diseases is also lower. This diet is rich in fruits, vegetables, fish, virgin olive oil and less saturated fat dairy products. Fats are necessary for the production of hormones and the absorption of the fat-soluble vitamins A, D, E and K.

Olive oil gives food a taste and a feeling of fullness, so that you are less hungry. This eating habit reduces the risk of cardiovascular disease, reduces the risk of diabetes, extends lifespan and there are more healthy elderly people.

Avoid fast food and unhealthy trans fats

The old McDonald's farm is very different from today's McDonald's. Fast food is in fashion. Hamburgers and chicken burgers are often on the menu of hard-working people. It's no longer rare for someone to go to bed with a bag of chips.

Vegetable (unsaturated) fats are liquid at room temperature and can only be processed as a solid by the food industry. These fats are converted into partially hardened fat by a chemical process. Products that contain partially hydrogenated fats and trans fatty acids, such as margarine, French fries, cookies, coffee creamer powder, cakes, crackers and pizzas, are bad for the blood vessels.

That's how you become a hundred years old

- do not fall
- do not have an accident
- don't catch a cold
- do not choke, pneumonia most common cause of death for centenarians

This is how you stay healthy

- no cold
- do not smoke
- with clean indoor air, no birds or cage animals in the house
- wash your hands regularly
- a glass of wine
- do not drink alcohol every day
- eat more vegetable proteins
- eat less or no red meat
- Training for an hour three times a week

Burn belly fat

Fat burning only starts after 20 minutes of exercise. Until this time of 20 minutes, especially the carbohydrate reserves (the glycogen in the liver and muscles) are burned and you do not lose weight. That is why it is better to exercise for an hour three times a week (3 x 40 minutes of burning fat) than six times a week for half an hour (6 x 10 minutes of burning fat). It does not matter what the training consists of. This could be a brisk walk, a slow jog, or a session on a treadmill or rowing machine.

Easily make your own meals

Follow this simple advice to get rid of excess body fat and prevent disease. Only a little whole meal bread, pasta and rice to lose weight faster. Food should contain a lot of vegetables, but also sufficient vegetable proteins and fats (fish, olive oil, avocado, etc.).

- **Start with a plant-based diet, then the need for animal proteins and fats will gradually decrease**

In addition to diet, physical activity and extra intake of antioxidants can counteract DNA methylation, which contributes to aging.

Fruits Breakfast

Start with two glasses of water

Oatmeal flakes with broken linseed in soy yoghurt or soya light, almond milk or coconut milk, with fresh fruit. Such as strawberries, raspberries, apples, pear, tangerine, orange, melon etc. Cut the fruit into pieces to preserve the dietary fiber.

Soup at lunch

Make soup from vegetable stock. Think of tomato, vegetables, onion, pumpkin, mushrooms, broccoli soup. A delicious soup can be made from all vegetables. Salad with nuts, mushrooms, arugula, tomato, onion, garlic, green beans, kidney beans, chickpeas etc. Olive oil dressing. Sandwich with salad of tuna, salmon, shrimps, etc. or an omelet or hard-boiled egg.

Omega-3 rich fish such as salmon, herring, mackerel and shellfish such as mussels are much healthier than red meat.

- No sausages, only hard-boiled eggs, no meat (products) from the supermarket, no undercooked BBQ meat, less dairy products, no raw milk cheese.

Hot meal

Make the most of herbs and spices. For example, replace the meat you were used to with chickpeas, brown or white beans. Make a delicious chili sin carne or curry dish with cauliflower, broccoli and chickpeas.

No dessert

Avoid sugars and easily digestible carbohydrates if you are overweight. By sweetening, the liver chooses the path of least resistance (glycolysis) and supplies the required energy, glucose and stores excess body fat.

After a meal, drink black coffee with a little almond milk, green tea or ginger tea. Ginger tea can be made by cutting slices of ginger root and boiling them in boiling water. Alcohol and wine in moderation. The degradation product acetaldehyde is harmful to our DNA. Too much tartaric acid damages the esophagus and stomach.

The farm shop and local markets

Finally, it goes without saying that it is better for our own health to put an end to the overproduction of animals for slaughter, pigs, chickens, eggs and their export. A number of pig and chicken slaughterhouses can be closed if production is only for domestic use and no longer for export abroad. Farm shop and local markets are safer than large supermarkets. As farmers in France did, large chains like McDonalds, Burger King and Kentucky Fried Chicken are now better kept out.

All the better restaurants now have a vegetarian menu.

Vertical Farming, Multilayer Cultivation

The usual agriculture and horticulture relies on fertilizers, pesticides and new plant varieties. In a closed system, no pesticides are needed anymore.

Cultivation in two weeks, which takes 30 days in the open ground, with 95 percent less water consumption and less fertilizer.

Sunlight cannot be controlled. With the cooler LED lamps, which can contain all kinds of colors, engineers can develop a light recipe. They can choose the right combination of wavelengths and light intensities, position the LED lamps near the plants and opt for wider or narrower light beams. Changing the color of the light changes the smell, taste and even the vitamin content of tomatoes.

The reason for the higher yield compared to greenhouses and outdoor cultivation is that under LED lighting the whole plant can receive sufficient light all year round and long days. In the uncontrolled sunlight, part is also lost because one plant gets too much and the other too little. Light is also lost due to reflection and the falling of light on the ground.

Strawberries are sweeter and tastier when the leaves and fruit are extra lit. In climate-controlled rooms, lettuce, spinach, bok choy, dill or cabbage, strawberries, coriander and watercress grow in four layers.

In an average Dutch greenhouse, the lettuce yield is 60 kilos per square meter of floor per year. In the vertical shelves, a yield of 100 kilograms per square meter of shelf is achieved.

Dozens of vertical farms, also known as 'vegetable factories' or 'indoor farms', supply spinach, bok choy, dill or cabbage every day. In Miyagi, Japan, a Japanese plant physiologist ordered 17,500 LED lamps to be installed in a former Sony factory. This factory supplies 10,000 unsprayed heads of lettuce every day. In Singapore, Panasonic opened a fully automated indoor farm for 81 tons of vegetables per year. Aero Farms in Newark, USA, opened its largest to date, a nine-foot warehouse that will supply 250 different unsprayed vegetables and herbs. Greenhouses are an area where machines are still surprisingly absent: picking crops such as tomatoes, peppers and strawberries has not yet been dominated by robotic hands on a large scale. All year round, armies of pickers move into the greenhouses, which usually have relatively low wages and long and hard working days. The Pick robot is already being used in Japan. Due to a lack of cheap labour, farmers there are in some cases content with robots that harvest even less than human pickers. Even if the robot only harvests sixty or seventy percent of all strawberries, the grower will earn more than if he hires relatively expensive pickers.

Agriculture on saline soils

- On Texel, they have succeeded in growing potatoes and vegetables on saline soil.
- Worldwide 1.5 billion hectares of agricultural land is threatened by salinization. In areas where salinization poses the greatest threat, this provides an opportunity to feed families independently.
- Can the salt potato save people from famine? Zilt Testing Company Texel has been researching which crops grow on saline soil since 2010. Many varieties do well.

Cultured meat

The stem cells from one gram of muscle tissue can be used to make approximately 10,000 kilos of meat. Mosa Meat (University of Maastricht) and the Dutch food giant Nutreco are food companies that want to enter the market in 2022 with meat from the laboratory. They make beef from the muscle cells of a cow. The cells multiply into trillions of cells from a small sample. This growth takes place in a bioreactor, comparable to the bioreactors in which beer and yoghurt are fermented. The cells are "brewed" in a liquid growth medium containing a mixture of fats, amino acids, carbohydrates, vitamins and minerals. The combination of ingredients in the liquid culture causes cells to differentiate into muscle, fat and connective tissue.

One cow tissue sample can make 800 million strands of muscle tissue (enough to make 80,000 quarter pounders). The difference is that it takes a cow about three years to develop enough meat to be slaughtered, but with cultured meat everything can be made in just a few weeks.

Global meat consumption fluctuates around 350 million tons per year. According to Mosa Meat, it is possible to produce 10 tons of meat from one tissue sample. That means 35 million tissue samples are needed to meet the current demand for animal food. Producing cultured meat therefore also has limitations and cannot replace current meat production.

References

Anttila TI, Koskela P, Leinonen M et al. (2003) Chlamydia pneumoniae Infection and the Risk of Female Early-Onset Lung Cancer. Int J Cancer:107,681-682

Bruu AL, Haukenes G, Aasen S, Grayston JT, Wang SP, Klausen OG, Myrmel H, Hasseltvedt V (1991) Chlamydia pneumoniae infections in Norway 1981-87 earlier diagnosed as ornithosis. Scand J Infect Dis 23(3):299-304

Chaturvedi AK et al. (2010) Chlamydia pneumoniae infection and risk for lung cancer. Cancer Epidemiol Biomarkers Prev 1498-1505

Chu DJ, Guo SG, Pan CF, Wang J, Du Y, Lu XF, Yu ZY (2012) An experimental model for induction of lung cancer in rats by Chlamydia pneumoniae. Asian Pac J Cancer Prev. 2012;13(6):2819-22

Chu DJ, Yao DE, Zhuang YF, Hong Y, Zhu XC, Fang ZR, Yu J and Yu ZY (2014) Azithromycin enhances the favorable results of paclitaxel and cisplatin in patients with advanced non-small cell lung cancer. Genet. Mol. Res. 13(2):2976-2805

Coggins CR (2001) A review of chronic inhalation studies with mainstream cigarette smoke, in hamsters, dogs, and nonhuman primates. Toxicol Pathol. 2001 Sep- Oct;29(5):550-7

Felini M, Preacely N, Shah N, Christopher A, Sarda V, Elfaramawi M, Sall M, Bangara S, Gandhi S, **Johnson ES** (2012) A case-cohort study of lung cancer in poultry and control workers: occupational findings. Occup Environ Med. 2012 Mar;69(3):191-7

Ferreri AJ, Ponzoni M, Guidoboni M et al. (2006) Bacteria-eradicating therapy with doxycycline in ocular adnexal MALT lymphoma: a multicenter prospective trial. J Natl Cancer Inst 98:1375– 1382.

Ferreri AJ, Dolcetti R, **Magnino** S ey al. (2007) A woman and her canary: a tale of chlamydiae and lymphomas. J Natl Cancer Inst. 2007 Sep 19;99(18):1418-9

Ferreri AJ, Govi S., Pasini E. et al. (2012) Chlamydophila psittaci eradication with doxycycline as first-line targeted therapy for ocular adnexae lymphoma: final results of an international phase II trial. J Clin Oncol Aug 20;30(24):2988

Gardiner AJ, Forey AB, Lee PN (1992) Avian exposure and bronchiogenic carcinoma. Br Med J 305 :989-992

Ger LP, Liou SH, Shen CV, Kao SJ, Chen KT (1992) Risk factors of lung cancer.J. Formos Med Assoc Sep; 91 Suppl 3:222-231

Jackson LA, Wang SF, Nazar-Stewart V, Grayston IT, Vaughan IL (2000) Association of Chlamydia pneumoniae immunoglobin A seropositivity and risk of lung cancer. Cancer Epidemiol Biomarkers Prev 9(11): 1263-1266

Johnson ES, Ndetan H, Lo KM (2010) Cancer mortality in poultry slaughtering / processing plant workers belonging a union pension fund. Environ Res 110(6):588-94

Johnson ES (2012), Choi Km. Lung cancer risk in workers in the meat and poultry industries - a review. Zoonoses Public Health 59(5):303-13

Jöckel KH, Pohlabeln H, Bromen K, Ahrens W, Jahn I (2002) Pet Birds and risk of lung cancer in North-Western Germany. Lung Cancer Jul;37(1)29-34

Kocazeybek B (2003) Chronic Chlamydophila pneumoniae infection in lung cancer, a risk factor: a case-control study. J Med Microbiol 52(8):721-6

Kohlmeier L, Arminger A, Bartolomeycik S, Bellach B, Rehm J, Thamm M (1992) Pet birds as an independent risk for lung cancer: Case-control study. Br Med J

Koyi H, Branden E, Gnarpe J, Gnarpe H, Arnholm B, Hillerdal G (1999) Chlamydia pneumoniae may be associated with lung cancer. Preliminary report on a seroepidemiological study. APMIS 107(9):828

Laurilla AL, Antilla T, Laara E, Bloigu A, Virtamo J, Albanes D, Leinonen M, Saikku P (1997) Serological evidence of an association between Chlamydia pneumoniae infection and lung cancer. Int J Cancer 20;74(1)1-34

Littman AJ Jackson LA, Vaughan TL (2005) Chlamydia pneumoniae and lung cancer: epidemiologic evidence. Cancer Epidemiol Biomarkers Prev. 14(4):773-8

Mather JP, Roberts PE, Pan Z et al. (2013) Isolation of cancer stem like cells from human adenosquamous carcinoma of the lung supports a monoclonal origin from a multipotential tissue stem cell. PLoS One 4;8(12)

Zhan P, Suo LJ, Qian Q, Shen XK, Qiu LX, Yu LK, Song Y (2011). Chlamydia pneumoniae infection and lung cancer risk: a meta-analysis. Eur J Cancer Mar;47(5):742-7

Publications and Books

Holst PAJ (1984) Bronchial carcinoma in bird keepers: an investigation in a general medical practice on a possible common relation. Ned Tijdschr Geneeskd 128:899-902

Holst PAJ, Kromhout D, Brand R (1988) Pet birds as an independent risk for lung cancer.

Br Med J 297:1319-1321

Holst PAJ (1997) Risk of lung cancer needs to be studied in younger patients, who keep and breed pet birds. Br Med J (1997) 314, 1353

Holst, PAJ (1991), Birdkeeping as a Source of Lung Cancer and Other Human Diseases. A Need for Higher Hygienic Standards.

Springer-Verlag ISBN 3-540-53555-1, Berlin/Heidelberg

Springer-Verlag ISBN 3-387-53555-1, New York

Kohlmeier L, Arminger A, Bartolomeycik S et al.(1992)

Pet birds as an independent risk for lung cancer: Case-control study. Br Med J 305:986-989

Chu DJ, Guo SG, Pan CF, Wang J, Du Y, Lu XF, Yu ZY (2012) An experimental model for induction of lung cancer in rats by Chlamydia pneumoniae. Asian Pac J Cancer Prev. 2012;13(6):2819-22.

Chu DJ, Yao DE, Zhuang YF, Hong Y, Zhu XC, Fang ZR, Yu J and Yu ZY (2014) Azithromycin enhances the favorable results of paclitaxel and cisplatin in patients with advanced non-small cell lung cancer. Genet. Mol. Res. 13(2):2976-2805

Holst, PAJ (2014) The Last Chimpanzee, somewhere in the 21st century, the last chimpanzee will die.

Paperback ISBN 978-94-02124-8-4

Holst, PAJ (2015) Plant-Based food is your Best Medicine

106 pages ISBN 9789082210569

Holst, PAJ (2016) Common Cancers are Zoonoses

197 pages ISBN 978-90-824963-3-8

Holst, PAJ (2016) Increase in Cancer is a Recent Event

ISBN 978-90-824963-0-7

Holst, PAJ (2016) PREVENTION IS BETTER THAN CURE

E-book 978-90-824963-2-1

Hamers, RWG (2017) De tijd van de Apocalyps

Paperback 270 pages ISBN 978-9463426664

Holst, PAJ (2019) Stop the Meatballs

Paperback 131 pages ISBN 978-1797658926

Holst, PAJ (2019) Our Inheritance from the Great Apes

Paperback. 146 pages ISBN 978-1081342159

Holst, PAJ (2019) Canimalism, 119 pages

Paperback 120 pages ISBN 978-1694357762

Hardcover 120 pages ISBN 978-9402198577

Holst, PAJ (2021) Save Our Selves en bescherm onze planeet Aarde

Paperback 143 pages ISBN 978-9403642239

Books consulted

A lot of scientific evidence has been published and also written in several books on health benefits of plant foods for various diseases in later life. Switching to strictly plant foods is not easy. The image of the classical vegetarian complicates this transition. But read more about the health benefits to be gained.

The China Study

Detailed study on the link between diet and heart disease, diabetes and cancer. The report also examines the source of confusing nutrition information from powerful lobbies, government agencies and opportunistic scientists. The China study observed whether there were association patterns for different diet, lifestyle and disease across 65 provinces, 130 villages and their families. This study in China and Taiwan also examined the virus hepatitis B (HBV), which causes primary liver cancer, a leading cause of death in Africa and Asia. They collected data on the prevalence of antibodies and antigens, mortality rates from multiple diseases, and many dietary risk factors. The prevalence of HBV antibodies was strongly correlated with vegetable consumption, dietary fiber and vegetable protein. In short, more plant food consumption was associated with more antibodies and an improved immune response.

Campbell TC, Campbell TM (2006) The China Study. electronic book Apple

How Not to Die

The vast majority of premature deaths can be prevented through simple diet and lifestyle changes. In How Not to Die, Dr. Michael Greger, the internationally renowned nutritionist, physician, and founder of NutritionFacts.org, lists the top 15 causes of premature death in America: heart disease, various cancers, diabetes, Parkinson's, high blood pressure, and more, and explains how nutritional and lifestyle interventions can trump prescription pills and pharmaceutical and surgical approaches, allowing us to live healthier lives.

Gene Stone & Michael Greger MD

Proteinaholic, how Our Obsession with Meat is Killing Us and What We Can Do About it

Whether you go to a doctor, nutritionist or trainer, they all recommend eating more protein. Foods, drinks and supplements are packed with extra protein. Many people use protein for weight management, to gain or lose weight, while others think it gives them more energy and is essential for a longer and healthier life. dr. Garth Davis, a weight loss expert, wonders: Does all this protein make us healthier? The answer, he emphatically states, is NO. Too much protein actually makes us sick, fat and tired, according to Dr. Davis. The healthiest countries in the world eat far less protein than we do and yet we have an entire nation that is getting sicker by the day.

Garth Davis MD & Howard Jacobson

Guns, Germs, and Steel: The Fates of Human Societies

Guns, Germs and Steel attempt to answer the greatest question in human history after the Ice Age: why Eurasian peoples, instead of peoples from other continents, became the ones who developed the ingredients of power (guns, germs and steel) and all over the world. to expand the world. Africans had a huge head start, as Africa is the continent with by far the longest history of human habitation. North America is a large fertile continent, as a result of which it supports the richest and most productive nation today. Australia provides by far the earliest evidence for the human ability to bridge large water gaps, and some of the earliest widespread evidence for behaviorally modern humans. Why were the Indos nevertheless the ones who expanded? The reasons were continental differences in the available wild plant and animal species suitable for domestication, resulting in an earlier, more prolific set of domesticated animals in Eurasia.

The east/west axis of Eurasia facilitated the spread of these domesticated animals throughout Eurasia. Europeans were able to spread at the expense of other peoples by infecting them (usually unintentionally) with epidemic infectious diseases such as smallpox and measles, against which Europeans had developed some genetic resistance and acquired much immune (antibody-based) resistance through historical and respectively lifetime exposure, while unexposed non-European peoples had no such exposure, and therefore no such resistance. The exchange of major epidemic infectious diseases was one-sided, because most of those diseases in the temperate regions came to us humans through diseases of our domestic animals (such as cattle, pigs and chickens) with which our ancestors lived in close contact after those animal species were domesticated. But of the world's 14 species of valuable domesticated mammals, 13 were Eurasian, only one American, and not a single Australian. Hence, Indos ended up as disease carriers, and with much resistance to their own diseases.

Jared Diamond, 1997

Dead Zone, where the wild things were

A tour of some of the world's most iconic and endangered species and what we can do to save them. Climate change and habitat destruction are not the only culprits behind so many animals facing extinction. The impact of consumer demand for cheap meat is equally devastating. We are mistakenly led to believe that compacting livestock for factory farming and growing crops on vast, chemical-soaked prairies is a necessary evil, an efficient way to cater for an ever-expanding world population as the land remains free for wildlife. Our planet's resources are reaching breaking point.

Every year, large amounts of nitrogen from fertilizers and manure are distributed over agricultural land. No doubt it increases crop yield, but plants don't take it up completely, so more fertilizer and animal waste is added than the plants need. Only a fraction of what is applied to the soil ends up in the crops. The rest flows to our rivers. Nitrogen and phosphorus levels, dead organisms are increasing in the Gulf of Mexico, Rhone Delta, North Sea, Baltic and Adriatic Seas. The oxygen content in these coastal waters is falling.

Dead Zone takes us on a dazzling research journey around the world, focusing on a dozen iconic species and, in each case, looking at the role industrial agriculture plays in their plight.

Phillip Lymbery. 2017

David Attenborough – A Life on Our Planet

Hundreds of investigations around the world have confirmed that something is going on. The consequences will be more far-reaching than the pollution of soil and water in some countries. Ultimately, this can lead to the disruption and collapse of everything we rely on.

This is the tragedy of our time: the accelerating decline of our planet's biodiversity. We need overwhelming biodiversity for life on our planet to truly thrive. Only when billions of different individual organisms make the best use of all the resources and opportunities they encounter, and when millions of species live interconnected lives that are interconnected in such a way as to sustain each other, can the planet function smoothly.

The greater the biodiversity of our planet, the safer all life on Earth, including ourselves, will be. However, biodiversity is collapsing under the influence of our current way of life.

We live our pleasant lives in the shadow of a disaster of our own making. This catastrophe is caused by the very things that allow us to create an atmosphere of well-being. And it makes sense that we would continue with this until we have a compelling reason to stop, and an attractive alternative. The natural world is in decline. The evidence is compelling that it will lead to our destruction. There is another alternative to turn the tide if we act now. Part of the solution may lie in the Netherlands, one of the few countries explicitly mentioned in the film. Attenborough explains that it is a densely populated country, but still the world's second largest exporter of food. **Attenborough** shows how efficiently vegetables are grown in our country, in greenhouses and under artificial sunlight. He also points out how the Netherlands deals with seawater. One of the problems we face is flooding in cities as a result of climate change. The Netherlands can also learn a lot about this from the rest of the world.

Acknowledgments

For the candidate exam I did my pathology exam with the former prof.dr. A. de Minjer. My thesis on small cell lung cancer was discussed and de Minjer took me to the pottery museum where we stopped for a while for a preparation with lung carcinoma from a smoker. De Minjer pointed out that lung cancer and breast cancer would be the biggest challenges of medicine for years to come. More than fifty years later, this is still the case.

Due to the many consultations and home visits, ten lung cancer patients came to my attention in the general practice. Of these, six were bird keepers in the years before the diagnosis. After consultation with the former Professor F. de Waard of the RIVM, Department of Epidemiology, I set up a ten-year practical study and follow-up research. The statistical relationship was demonstrated, later confirmed in studies in Berlin and Glasgow.

I would like to thank the former Professor Zwart (Department of Veterinary Pathology, Department of Diseases of Special Animals, University of Utrecht) for his comments after the presentation of my research results. "Original ideas and observations are rare. They are especially valuable when verified in the field, critically evaluated, and supported by material independently collected by others."

Holst saw a possible link between the keeping of birds and the occurrence of lung cancer in members of the household where they are kept. He has pursued the idea in his private practice and has been tracking all patients for over 12 years. The data has been critically and statistically analyzed and supplemented with data and materials collected by lung specialists". Cancer research, especially lung cancer in humans, has involved a great deal of scientific input and has contributed significantly to understanding the many factors involved. A new aspect is presented in bird products, dispersed in the form of fine dust particles, inhaled deeply, causing irritation and contributing to local immune reactions in the lungs. Much later, in 2012, lab experiments proved the link between lung cancer and Chlamydia pneumonia infection (**Chu DJ 2012 & 2014**).

Dr. Michael Greger MD was very predictive with his publication on the origin of zoonoses.

Greger, M. (2007). The human/animal interface: emergence and resurgence of zoonotic infectious diseases. Critical Reviews in Microbiology, 33(4), 243-299.

Biography

Dr. Peter Holst worked until 1984 as a general practitioner in the region of The Hague, the Netherlands. In 1970, as a starting general practitioner, he saw a young girl with severe Chlamydia pneumonia. She recovered after treatment with an antibiotic. Because she kept a caged parakeet in her bedroom, he assumed that "the presence of a caged bird at home could be responsible for a more serious illness." In his practice he also treated a 17-year-old boy with an osteosarcoma in his upper leg, from which he died. As a hobby he kept and bred about 100 tropical birds in a basement room.

Due to the many consultations and home visits, ten lung cancer patients came to his attention in the following years. Of these, six were bird keepers in the years before the diagnosis. After consultation with a professor of epidemiology, he started a ten-year practical study. He conducted his research at Utrecht University. Unique because of the combination of veterinary medicine and human medicine. Diseases that can be passed from animals to humans (zoonosis) are becoming increasingly common. With the support of the **Dutch Prevention Fund**, he conducted research into new cancer cases in his own practice (ten-year field study). The results have been published in the Dutch Journal of Medicine (Holst 1984). After this, he started a case-control study of all newly diagnosed lung cancer patients in all hospitals in The Hague with the help of their own lung specialists. The results of these studies have been published in the British Medical Journal (**Holst, Kromhout & Brand 1988**). Keeping birds and breeding birds have been shown to pose a risk of lung cancer. In the study of lung cancer patients in The Hague, in addition to the risk of keeping birds in the house, especially breeding birds, they also found a reduced intake of vitamin C in fresh fruit and vegetables.

Together with the Netherlands Organization for Applied Scientific Research (TNO Delft), he then carried out dust measurements in the homes of bird keepers.

This research led to his PhD at the University of Utrecht in 1987 on the relationship he demonstrated between breeding and keeping birds indoors and lung cancer. He defended the hypothesis that lung cancer in bird keepers and breeders is the result of persistent infection of the deeper basal cells in the airways. These basal cells are still multi-potent and will not die if the cell is infected with a bacterium such as Chlamydia that can only reproduce in a living host cell. His promoters were Prof. F. de Waard, epidemiologist of the RIVM, Prof.dr. P. Zwart, head of the Faculty of Veterinary Medicine at Utrecht University and D. Kromhout, nutritional epidemiologist.

General practitioner examinations and dust measurements with TNO were supported by the Dutch Prevention Fund.

Holst specialized in occupational health and environmental health from 1984 at the Netherlands Institute for Preventive Medicine (NIPG-TNO Leiden). Since its foundation in 1991, he has been a member of the International Society of Indoor Air Quality (ISIAQ). He has published in several medical journals and has written books on indoor hygiene and preventive medicine.

After his retirement in 2005 he traveled a lot. Born in Zeeland, land in the sea, he is attracted to the wide world and the inlet. After more than 20 cruises, he has now crossed all the oceans several times. The volcanic islands in the Pacific Ocean are very impressive. All first life forms originated here and have spread to America, Africa, Europe, Asia and Australia. Fish, amphibians, birds, dinosaurs and mammals originate from the primordial soup of the Pacific Ocean. He noted that there were no great apes in America and Australia. The oldest human civilizations are found in the Far East. Recent DNA research has shown that 70,000 years ago the Aborigines reached Australia from East Africa via Antarctica.

In 2012, a lab experiment proved the link between lung cancer and Chlamydia pneumonia infection.

His interest in the link between tropical bird breeding and cancer has expanded to the health risks of the intensive rearing of poultry, pigs and cattle for consumption. Artificial breeding of livestock has increased dramatically since the 1950s. Over the past fifty years, our diet has become increasingly unnatural. More meat products from animals, bred solely for consumption, cause more chronic diseases. An increase that keeps pace with the recent rise in cancer mortality

www.ingramcontent.com/pod-product-compliance
Ingram Content Group UK Ltd.
Pitfield, Milton Keynes, MK11 3LW, UK
UKHW040032200726
13854UKWH00001B/482